AF786711

Cobalt Investing Outlook

Roggie Clark

Learn How To Invest in Gold Bars With This Book

Whether you're new to investing or seeking to diversify your portfolio, this beginner-friendly guide will give you the knowledge and confidence to invest in gold, and other precious metals.

https://www.amazon.com/Ultimate-Investing-Buying-Guide-Beginners-ebook/dp/B0F1KWWDXG

Learn How To Invest in Silver With This Book

Whether you're approaching retirement or already enjoying it, protecting and growing your wealth is critical. The Ultimate Silver Investing & Buying Guide For Beginners is your clear, no-nonsense guide to incorporating silver into your wealth-building strategy.

https://www.amazon.com/Ultimate-Silver-Investing-Buying-Beginners-ebook

Learn Macro and Micro Economics, With This One Book

Whether you're a student, a professional, or simply someone eager to understand the world around you, this book is your perfect introduction to the fundamentals of micro and macroeconomics.

https://www.amazon.com/Macro-Micro-Economics-Made-Beginners-ebook/dp/B0DPHTFVQG

Learn How To Invest in Copper With This Book

Are you ready to take control of your financial future and explore the world of copper investing? Whether you're new to investing or seeking to diversify your portfolio, this beginner-friendly guide will give you the knowledge and confidence to invest in copper.

https://www.amazon.com/Ultimate-Copper-Investing-Buying-Beginners-ebook/dp/B0F892PBSJ

Copyright

Title: Cobalt Investing Outlook
Author: Roggie Clark

Copyright © 2025 by Roggie Mullis Clark III

All rights reserved. No part of this book may be reproduced, stored in a retrieval system, or transmitted in any form or by any means, electronic, mechanical, photocopying, recording, scanning, or otherwise, without the prior written permission of the author and publisher except in the case of brief quotations embodied in critical reviews and certain other noncommercial uses permitted by copyright law.

All trademarks, trade names, or logos mentioned or used in this book are the property of their respective owners and are used for identification purposes only.

Cover Design: Roggie Clark
Editing: Renee' Clark

Printed in the United States of America

Disclaimer

The views and opinions expressed in this book are those of the author and do not necessarily reflect the official policy or position of any individual, company, or organization cited or referenced in the text.

The author is not affiliated with any of the individuals, companies, or organizations mentioned in the text, nor does the author receive any form of compensation for citing or referencing them. Some sample scenarios in this text are fictitious. Any similarity to actual persons, living or dead, is purely coincidental.

The professional information (strategies, tips, and concepts) in this text is provided on an "as is" basis, without warranties of any kind, express or implied. While every effort has been made to ensure the accuracy of the information contained herein, neither the author nor the publisher makes any representations or warranties regarding its completeness, reliability, or applicability. The information is for educational and informational purposes only and should not be interpreted as professional, legal, financial, or business advice.

This book does **not** constitute financial advice, investment guidance, or legal counsel. Any financial, business, or investment decisions made based on this book are done **at your own risk**. You should always conduct your own research and consult a **certified financial**

advisor, legal professional, or other qualified expert before making any decisions that could impact your finances, business, or legal standing.

Neither the author nor the publisher guarantees any specific results or outcomes. By reading this book, you acknowledge and agree that the author and publisher shall not be held liable for any direct, indirect, incidental, or consequential damages, financial losses, or harm arising from the use or misuse of the information contained in this text. You assume full responsibility for any actions taken based on the content of this book.

Table of Contents

Introduction

Cobalt is a hard, bluish metal that has become increasingly important in modern energy and technology systems. While cobalt has been used for centuries as a coloring agent in glass and ceramics, its modern value comes from its role in rechargeable batteries and high strength alloys. The metal supports technologies ranging from smartphones to electric vehicles, making it a central part of the shift toward cleaner and more efficient energy solutions.

As industries push for longer battery life, better energy storage, and improved performance, demand for cobalt has grown significantly. This book provides a simple, top level overview designed to help investors understand why cobalt is becoming more visible in discussions about strategic materials and future energy systems.

Why this metal is getting attention

Cobalt is gaining attention largely because of its essential role in lithium ion batteries, the same batteries that power electric vehicles, laptops, and many other portable devices. These batteries rely on cobalt to maintain stability, manage heat, and extend their overall lifespan.

As the world moves toward electrified transportation and expanded renewable energy storage, materials that support reliable battery technology have become more important. Cobalt also plays a role in aerospace, manufacturing, magnets, and specialized engineering where strength and durability are critical. These rising applications help explain why conversations surrounding cobalt have become more common in recent years.

Early interest and why it matters

Early interest in cobalt reflects the growing importance of battery technology and the materials that make it possible. As countries invest in electric transportation and renewable energy systems, the metals that support these transitions naturally become part of the discussion.

Cobalt's use in high performance energy storage makes it a key material for industries focused on efficiency and technological growth. Understanding why cobalt matters helps investors follow broader trends in energy policy, manufacturing, and resource planning. This awareness becomes increasingly valuable as more companies and governments plan for a future built on reliable and sustainable energy solutions.

Last thoughts

This introduction highlights how cobalt has moved from a relatively niche metal to a central player in battery technology and modern energy systems. Even though many people do not interact with cobalt directly, its influence is present in many of the devices they use every day.

Recognizing cobalt's growing importance helps investors make sense of shifting technological and industrial priorities. This book aims to offer a simple perspective that supports early awareness rather than technical guidance. With this foundation in place, the next chapter explores what cobalt is and what makes it unique among metals.

Chapter 1: What the metal is

Cobalt is a hard, brittle metal with a natural silvery gray appearance that often shows a faint blue tint. It is typically found in combination with other elements rather than in pure form, which means it must be extracted during the processing of other minerals.

While cobalt has long been valued for its ability to create rich blue pigments, its modern significance comes from properties that support high strength alloys, advanced batteries, and magnetic materials. Cobalt's unique combination of stability, hardness, and heat resistance sets it apart from many other metals. Understanding what cobalt is provides a foundation for recognizing why it has become a critical industrial material.

Basic identity and key characteristics

Cobalt belongs to the transition metal group, and it is most commonly obtained as a byproduct of mining nickel and copper deposits. It is solid and dense, and it maintains its strength even at high temperatures, which gives it special value in industrial processes that require durability and heat resistance. Cobalt forms stable alloys that do not break down easily, so it is used in jet engines, gas turbines, and other mechanical systems that operate under extreme stress.

One of cobalt's most famous characteristics is its natural oxide, which produces a strong and vibrant blue color when heated. This oxide was ground into pigments for thousands of years, long before cobalt was understood scientifically. Ancient craftspeople in Persia, Egypt, and China used cobalt rich minerals to color pottery, tiles, and glass, creating the bright cobalt blue shades still seen in historic artifacts.

The pigment binds well to ceramic materials during firing, and it does not fade easily, so it became a reliable source of long lasting color for artwork, architectural decorations, and fine porcelain. Even today, artists and ceramic manufacturers continue to use cobalt oxide because it delivers clean, intense shades of blue that remain stable under high heat.

Although this artistic use still exists, modern industries rely far more on cobalt for its mechanical and chemical properties. Its identity as both an industrial and technological metal helps explain why it remains in steady demand, especially as global interest grows in electric vehicles, clean energy storage, and high performance manufacturing.

What makes it different

What sets cobalt apart from other metals is its unique combination of thermal stability, mechanical strength, and electrochemical performance. Many common metals such as aluminum, iron, and zinc lose strength or oxidize rapidly at high temperatures, but cobalt remains stable and maintains its structural integrity even under intense heat.

This resilience allows it to function in environments where metals like copper or nickel would soften, deform, or corrode. In battery chemistry, cobalt also outperforms metals such as manganese and iron because it helps stabilize the cathode structure, improves energy density, and reduces the risk of thermal runaway. Few metals support lithium ion battery chemistry as effectively as cobalt, which is why it plays such a key role in high energy and long cycle life battery designs.

Its ability to withstand high stress and high heat also positions it as a standout material for aerospace and industrial applications. Metals like titanium and stainless steel are strong, but cobalt based superalloys resist creep, cracking, and fatigue under extreme conditions better than many competing materials.

While alternatives to cobalt are being researched, including nickel rich battery cathodes and manganese based chemistries, its performance advantages remain difficult to match across several major industries. These differences highlight why cobalt continues to be viewed as a strategically important resource and why demand for it remains steady even as new technologies evolve.

Last thoughts

Understanding the basic identity and characteristics of cobalt provides a clear view of why the metal has become so important. Its combination of strength, stability, and performance sets it apart from more common materials. As industries continue to evolve, cobalt's unique properties keep it at the center of many conversations.

This chapter establishes the groundwork needed to understand its modern uses. With these fundamentals in place, the next chapter explores in more detail in which industries cobalt is used and why those applications matter so much for advanced and strategic technologies.

Chapter 2: Advanced uses and applications

Cobalt is used across a wide range of industries, from battery manufacturing to aerospace engineering, due to its ability to support high performance materials. As stated before, its unique combination of stability, conductivity, and durability gives it advantages in technologies that require long lifespan and reliable operation.

Although most people interact with cobalt through devices powered by lithium ion batteries, the metal plays a much broader role behind the scenes of some of the world's most advanced technologies. This chapter highlights in greater detail the main areas where cobalt is used and why those applications matter.

Strategic or high tech uses

As stated before, cobalt is used in jet engine components and superalloys where long term heat resistance is critical. Cobalt based superalloys retain their strength at temperatures that would weaken or melt many other metals, which makes them essential in turbine blades, combustion chambers, and exhaust systems. Cobalt also helps engines run hotter and more efficiently, which increases fuel performance and extends engine life.

In defense applications, cobalt alloys appear in military jet engines, missile systems, heat resistant armor components, and guidance technologies that must function under high thermal and mechanical loads. Its stability and reliability make it a strategic material for national security and aerospace manufacturing.

Certain electronics and medical devices also rely on cobalt for its magnetic properties and specialized chemical behavior. Cobalt is a key element in high strength permanent magnets used in radar equipment, satellite systems, electric motors, robotics, and precision sensors. In the medical field, cobalt chromium alloys are used in implants such as artificial hips and knees because they resist corrosion and wear. These high tech uses place cobalt at the center of fast growing industries.

Why these uses are important

The importance of cobalt's applications becomes clear when considering how many modern technologies rely on reliable and efficient energy storage. Electric vehicles, renewable energy storage systems, and portable electronics all depend on batteries that perform consistently. Cobalt's ability to support battery stability plays a major role in meeting these demands.

Additionally, industries that rely on high strength alloys and heat resistant materials require cobalt for systems that must function under extreme conditions. These uses emphasize why cobalt continues to be essential in conversations about technological growth and energy transition.

Last thoughts

Cobalt's widespread applications show how deeply the metal is integrated into essential industries. Its influence stretches from energy storage to aerospace, electronics, and medical making it a material that supports both everyday and high level technologies.

As global industries continue to grow, the demand for cobalt is likely to remain strong. Understanding these uses helps investors better understand why the metal receives so much attention. With this foundation in place, the next chapter explores where cobalt comes from and how supply factors shape its role in today's technology landscape.

Chapter 3: Supply and availability

Because cobalt rarely forms large, concentrated deposits of its own, its production depends heavily on the success and stability of these larger mining industries. This creates a supply environment where availability can shift due to factors unrelated to cobalt directly.

Additionally, cobalt production is heavily concentrated in a small number of countries, which raises long term questions about access, reliability, and geopolitical influence. This chapter explores where cobalt comes from, how much is produced, and what challenges shape its global availability.

Where the metal comes from

Most of the world's cobalt supply comes from the Democratic Republic of the Congo. Significant amounts of cobalt are also produced in countries like Russia, Australia, Canada, and the Philippines, though their contributions are smaller in comparison. Because cobalt is sourced as a byproduct, countries with large nickel and copper operations naturally play a larger role in supplying the metal.

Refining capacity is more globally distributed, with processing taking place in regions such as China, Finland,

and South Korea. Understanding where cobalt originates helps clarify why the metal is often discussed in the context of localizing mining, processing, and refining operations, especially for Western nations.

Cobalt mining at scale

Each year, cobalt production totals only a fraction of what more common metals produce, which contributes to its classification as a critical material. China processes a large portion of the world's cobalt into refined products, giving it major influence over global availability.

Even though the Democratic Republic of the Congo supplies most of the raw cobalt ore, China dominates the midstream and downstream parts of the supply chain through refining, chemical processing, and component manufacturing. This creates a situation where mining and refining occur in different regions, and the global market depends heavily on stable trade relationships and predictable export policies. Other countries are attempting to expand their refining capabilities, but China still holds a significant lead in processing capacity.

Last thoughts

Understanding cobalt's supply and availability provides important context for its role in global technology and energy systems. The metal's dependence on other mining industries, combined with concentrated production regions, adds complexity to long term supply planning.

As demand increases from electric vehicles and energy storage industries, these supply factors become even more significant. This chapter clarifies why cobalt is often treated as a strategic and carefully monitored resource. With supply considerations established, the next chapter explores why more people are talking about cobalt and what is driving its rising visibility.

Chapter 4: How investors may be positioning themselves

As cobalt becomes more important to energy and strategic technology, early observers are paying closer attention to how the metal fits into long term trends. While cobalt does not have a straightforward consumer investment market, people who follow emerging materials often analyze the industries and supply chains connected to it.

This high level approach focuses on understanding broader signals rather than looking for direct investment opportunities. Because cobalt influences major technologies such as electric vehicles and battery storage, awareness matters more than speculation. This chapter highlights general ways people may be monitoring cobalt as its relevance expands.

High level exposure categories

Observers who want to understand cobalt's influence often follow companies involved in battery production, electric vehicle manufacturing, and mining industries. These sectors rely heavily on cobalt, making them useful indicators of broader demand trends.

People also pay attention to copper and nickel mining operations, since cobalt is commonly produced as a byproduct of these activities. Energy storage companies, renewable technology manufacturers, and research groups working on next generation batteries can also offer insight into cobalt related developments. These indirect exposure categories help people watch cobalt's broader ecosystem without relying on formal investment mechanisms.

What early adopters watch

Early adopters monitor supply chain announcements, new mining projects, and changes in battery chemistry research to understand cobalt's potential direction. Updates involving electric vehicle adoption rates or renewable energy initiatives can signal long term demand trends.

Observers may also watch geopolitical developments, especially discussions about responsible sourcing and supply diversification. Battery innovation efforts, including attempts to reduce cobalt usage, are another area that early followers track closely.

Last thoughts

This chapter shows how people may be following cobalt by watching high level trends across energy, technology, and manufacturing. Observers rely on signals such as electric vehicle growth, battery advancements, and supply chain developments rather than traditional investment paths.

Cobalt's influence on major industries makes it a material worth monitoring from this conceptual viewpoint. The focus here remains on understanding rather than speculation. With these ideas in place, the next chapter explores what the future could look like for cobalt as clean energy and technology continue to evolve.

Chapter 5: What the future could look like

As countries move toward cleaner energy goals, the materials that support high performance storage solutions become increasingly important. At the same time, supply concerns, technological innovation, and efforts to diversify materials are shaping conversations about cobalt's long term role. While this book does not aim to predict outcomes, understanding potential directions helps frame how cobalt may fit into future developments.

Potential new uses

As research continues, cobalt may find expanded applications in next generation battery chemistries, advanced catalysts, and high temperature materials. Scientists are developing solid state batteries, sodium ion batteries, and lithium sulfur batteries, and some of these designs may still benefit from cobalt's ability to stabilize cathodes, increase energy density, and improve thermal safety.

Performance focused electric vehicle platforms may continue to rely on cobalt intensive battery designs, especially for long range, high acceleration models that require stable, heat tolerant materials. Heavy duty electric trucks, aircraft batteries, and grid scale storage systems might also use cobalt rich formulations if they deliver superior durability.

Aerospace companies could increase their use of cobalt based alloys as engines and turbines evolve to support new energy technologies. Future propulsion systems, including hybrid electric engines, hydrogen fueled turbines, and supersonic or hypersonic aircraft designs, may demand materials that tolerate even higher stress and temperature levels. Cobalt superalloys provide strength and resistance to oxidation that allow engineers to push performance thresholds further than many nickel or titanium based alloys alone.

Cobalt may also play a growing role in advanced chemical and energy systems. Researchers are investigating cobalt catalysts for carbon capture processes, synthetic fuel production, ammonia cracking, and hydrogen generation.

Cobalt's catalytic behavior helps accelerate reactions that are difficult for metals like iron or copper to support, making it valuable for cleaner and more efficient energy conversion technologies. Some early stage studies are exploring cobalt based materials for water splitting reactions, which could support future hydrogen economies.

Certain medical devices, sensors, and magnetic materials may also incorporate cobalt due to its unique properties. Cobalt chromium alloys are being studied for next generation orthopedic implants that offer better wear resistance and longer lifespans. Emerging fields such as quantum computing, neuromorphic electronics, and high density data storage may also create new demand for cobalt.

Some experimental memory devices rely on cobalt for its magnetic switching properties, and thin film cobalt layers are being tested in spintronic devices that use electron spin instead of electrical charge to process information. These potential new uses illustrate how cobalt may continue to support innovations that have not yet reached wide adoption.

Possible market expansion

Efforts to secure a reliable cobalt supply may encourage the development of new mining operations outside of the dominant producing regions. Countries interested in energy independence and sustainable sourcing are studying geological surveys, investing in exploratory projects, and creating incentives for domestic mining to reduce reliance on imports.

Many governments are also strengthening international partnerships to stabilize supply chains. Recent trade agreements and cooperative frameworks involving the Democratic Republic of the Congo reflect this strategy, since the DRC remains the world's largest source of raw cobalt. These deals aim to improve transparency, support responsible mining practices, and provide long term access to cobalt for industries that depend on it.

Countries focused on energy independence and sustainable sourcing are exploring opportunities to diversify production and refining capabilities. New refining facilities in regions such as North America, Europe, and parts of Asia are intended to reduce the concentration of processing power in a small number of countries. Battery recycling is also likely to play a growing role in supplementing primary cobalt supply, especially as electric vehicle adoption accelerates and the first major wave of end of life batteries enters the recycling stream.

Improved refining processes and expanded cooperation between mining companies, technology manufacturers, and national governments may also help stabilize the market. Shared investments in cleaner extraction methods, digital tracking systems, and supply chain monitoring can raise production quality and strengthen reliability.

These developments could broaden access to cobalt, reduce dependence on a small number of suppliers, and create a more resilient global supply chain that supports long-term growth in energy storage and advanced manufacturing.

Last thoughts

Looking ahead, cobalt's future is shaped by innovation, energy policy, and global demand for reliable battery technology. Some research efforts aim to reduce the amount of cobalt in batteries, but many alternatives still struggle to match cobalt's performance characteristics. Its unique properties position it well for industries that value safety, performance, and long lifespan in energy systems.

While the exact path forward is uncertain, the trends outlined in this chapter show why cobalt remains a material worth watching. The possibilities for expanded production, new applications, and improved recycling suggest that cobalt could continue playing a major role in the energy transition. With these ideas established, the next chapter provides a closing summary and reinforces the value of early awareness.

Final thoughts

This book has provided a clear and simple overview of cobalt and its growing importance in modern energy and technology systems. From its role in rechargeable batteries to its use in high strength alloys and advanced electronics, cobalt supports many of the innovations shaping today's world.

Each chapter offered a different piece of the larger picture, helping investors understand what cobalt is, where it comes from, how it is used, and why it continues to be discussed across industries.

The goal was not to offer investment advice but to build awareness around a metal that quietly powers significant portions of the global energy transition. With this foundation, investors can better follow future developments involving cobalt as technology and policy continue to evolve.

The goal is awareness

The purpose of this book is to help investors gain a straightforward understanding of why cobalt matters in modern industries. Its role in stabilizing battery chemistry, supporting electric transportation, and enabling advanced engineering makes it a material worth recognizing. Awareness of cobalt's influence helps investors interpret changes in global manufacturing, energy strategies, and supply chain planning.

Understanding these connections provides valuable context for following long term discussions involving responsible sourcing, innovation, and resource security. This top level awareness is especially useful as conversations about cobalt continue to grow.

Staying informed over time

Staying informed about cobalt means paying attention to developments in battery technology, electric vehicle adoption, and global supply diversification. Updates related to mining reforms, environmental standards, and recycling efforts can also offer insight into cobalt's future.

Observing these broader trends helps investors understand how cobalt fits into the evolving technological landscape without needing to track formal markets. Keeping an eye on these signals supports continued awareness of a metal that plays a central role in the global shift toward cleaner energy. With this perspective, investors are better prepared to appreciate future conversations involving cobalt and other strategic materials.

About the Author

Roggie "Raj" Clark is an entrepreneur and a 10-year marketing professional with a proven track record in e-commerce and digital business growth.

Having built and managed successful online stores with thousands of sales and consulted for numerous marketing clients, Roggie has mastered the intricacies of online marketing, branding, and consumer behavior. He has also written several popular books on digital marketing, SEO (search engine optimization), and programming (HTML, CSS, and JavaScript).

As a former marketing manager at a fintech company—a role he secured through his own marketing and financial expertise—Roggie gained firsthand experience in financial strategy and wealth-building.

This journey inspired him to share insights with amateur investors, helping them save money, make smarter financial decisions, and build long-term wealth. Raj currently lives in Maryland with his sister and niece.

Printed by Libri Plureos GmbH in Hamburg,
Germany

9 798295 562433